ULTIMATE SUPERCARS

CHEVROLET COPO CAMARO

By Joanne Mattern

Kaleidoscope
Minneapolis, MN

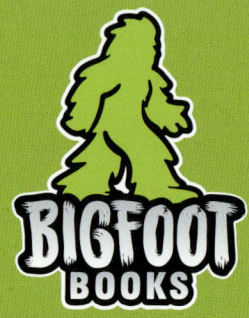

The Quest for Discovery Never Ends

....

This edition first published in 2023 by Kaleidoscope Publishing, Inc.

No part of this publication may be reproduced in whole or in part without written permission of the publisher.

For information regarding permission, write to
Kaleidoscope Publishing, Inc.
6012 Blue Circle Drive
Minnetonka, MN 55343

Library of Congress Control Number
2022937985

ISBN
978-1-64519-610-5 (library bound)
978-1-64519-680-8 (ebook)

Text copyright © 2023 by Kaleidoscope Publishing, Inc. All-Star Sports, Bigfoot Books, and associated logos are trademarks and/or registered trademarks of Kaleidoscope Publishing, Inc.

Printed in the United States of America.

Bigfoot lurks within one of the images in this book. It's up to you to find him!

TABLE OF CONTENTS

Chapter 1: What a Ride! 4

Chapter 2: Made to Order 12

Chapter 3: The Big Block 18

Chapter 4: Around the Track 24

Beyond the Book 28

Research Ninja 29

Further Resources 30

Glossary 31

Index 32

Photo Credits 32

About the Author 32

Chapter 1
What a Ride!

Marco wiggled with excitement as his Uncle Chris pulled into the parking lot of the **drag-racing** track. The boy unbuckled his seat belt and followed his uncle into the building. "Where is your new car, Uncle Chris?" he asked. "I can't wait to see it."

"There it is," Uncle Chris said. "Say 'hello' to my new 2022 Chevy Copo Camaro."

Marco walked around the car. It looked fast. His uncle opened the shiny, **carbon-fiber** hood to show him the engine. Marco could hardly believe his eyes. The engine was huge! "This car has a 572-**cubic-inch** big-block **V8** engine," Uncle Chris said. "That gives the car a lot of power."

FUN FACT
The 2022 Copo Camaro can have three different engines. The 572 is the most powerful.

Marco could not wait to see the car in action. He stepped off the track and waited behind a safety railing. When his uncle started the engine, it *sounded* fast.

The Camaro nosed up to the starting line. The engine purred. Then it roared to life as his uncle touched the gas pedal. Marco covered his ears to block the screaming sound of the engine. But he could not take his eyes off the car.

Zoom! The back wheels spun as the Camaro powered forward. It passed Marco in a blur of color. Marco gasped as the front wheels lifted off the ground for a second.

Uncle Chris has told him that the Copo Camaro could go from 0 to 60 miles per hour in 5.4 seconds. Marco timed the car's run down the quarter-mile track. It took just 9.6 seconds for it to cross the finish line.

"How fast did you go?" Marco asked when Uncle Chris brought the car back to the starting line.

Uncle Chris grinned. "More than 142 miles per hour. What a ride!"

Marco watched his uncle drive the car into the garage. He could not wait until he was old enough to drive the Camaro himself.

NOT ON THE STREET

You will not see the Chevy Copo Camaro driving around your neighborhood. The car is not street legal. It is against the law to drive it. The Copo Camaro is just too fast to be safe on the road. This car can only be driven on a racetrack.

PARTS OF A
COPO CAMARO

carbon-fiber hood

Chevy bow tie

racing stripe

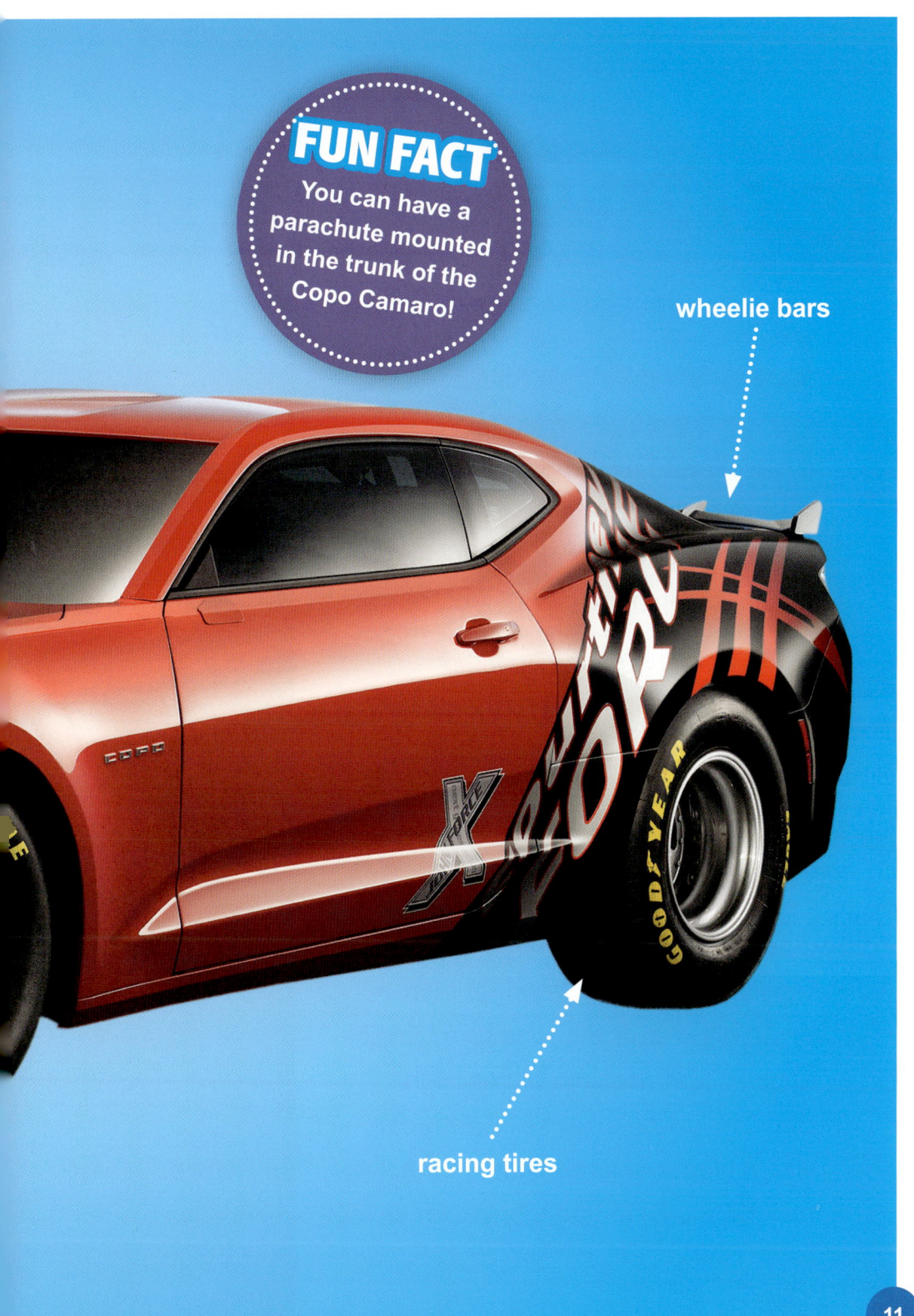

FUN FACT
You can have a parachute mounted in the trunk of the Copo Camaro!

wheelie bars

racing tires

Chapter 2
Made to Order

Chevy is a nickname for Chevrolet. The car company started in 1911. It was started by a race car driver named Louis Chevrolet and a man named William Durant. Durant started the car company General Motors.

Chevrolet soon became known for making cars that were fun to drive. In 1955, it introduced a small-block engine. This engine gave a lot of power to cars.

Fred Gibb sold Chevy cars. He also liked to race Camaros. However, regular Chevy cars were not fast enough to win at the racetrack. Their engines just were not big enough. But Chevy would not sell Camaros with a bigger engine. Then, in the mid-1960s, Gibb had a bright idea.

Chevy had a system called Central Office Production Order. It was called COPO for short. COPO let dealers order special parts for certain cars. Gibb and some friends at Chevy used COPO to order a ZL1 427-cubic-inch engine. He also ordered a special rear axle and special tires. He added more powerful brakes and a stronger transmission. Now Gibb had a car that could do well on the racetrack.

The new car was the Copo 9560.

Fred Yenko was another Chevy dealer who liked to race. In 1967, he ordered cars and then changed their engines. He added a 427-cubic-inch engine from the Chevy Corvette to create the Copo 9561. Lots of people wanted Yenko's super-fast cars.

The first Copo Camaros hit the racetracks in 1969. However, only a few hundred were made. The cars are very rare.

Then, in 2012, Chevy decided to make new Copo Camaros. The 2012 Copo ZL1 had a choice of three V8 engines. Two of these engines were super-charged. They provided up to 580 **horsepower**. Chevy also offered other gear that drag-racing teams could add to the car.

The Copo Camaro got a boost in 2017. That's when Chevy added a 650-horsepower engine. The 2022 model of this car is even faster and stronger.

FUN FACT
People who buy a 2022 Copo Camaro can ask for a tour of the factory where the car is made.

WHERE IS THE CHEVY COPO CAMARO MADE?

Oxford, Michigan

SETTING LIMITS

For years, Chevy limited the number of Copo Camaros it made. Only 69 Copo Camaros were built in 1969. Between 2012 and 2022, Chevy built fewer than 700 of these cars. However, Chevy did not limit the number of 2022 Copo Camaros it made.

Chapter 3
The Big Block

The most amazing thing about the 2022 Chevy Copo Camaro is its engine. The engine is big. Really big.

Open the hood of the Copo Camaro, and take a look. The engine seems to barely fit under the hood. The car's engine measures 572 cubic inches. That is the largest engine ever put into a Copo Camaro. This big engine creates over 700 horsepower. That's enough to keep this car moving fast around the track.

Does the 572-cubic-inch engine seem too big? No worries! The car is also available with a super-charged 350-cubic-inch V8 engine. This engine provides 580 horsepower. Or there is a 427-cubic-inch engine with 470 horsepower. No matter what engine, this car will tear up the track!

The engine is made of aluminum, iron, and steel. It is powered by a fuel injection system. This system sprays the fuel and mixes it with air before it flows into the engine. The car's computer senses exactly how much fuel to release and how long it has to mix with air. This allows for the best mix of fuel and air to power the engine. The better the fuel mix, the better the engine runs.

The car has a three-speed automatic transmission built by ATI Racing Products. That makes it easy and fun to drive. Go ahead and floor the gas pedal. This car is ready to respond!

THE COPO CAMARO IN DETAIL

COST: $105,500 basic model

Height: 53.1 inches (1349 mm)

Width: 74.7 inches (1897 mm)

LENGTH: 188.3 inches (4,783 mm)

WEIGHT: 3,175 pounds (1,440 kg)

TOP SPEED: 175 mph (282 kph)

TIME FROM 0 to 60 mph: 5.4 seconds

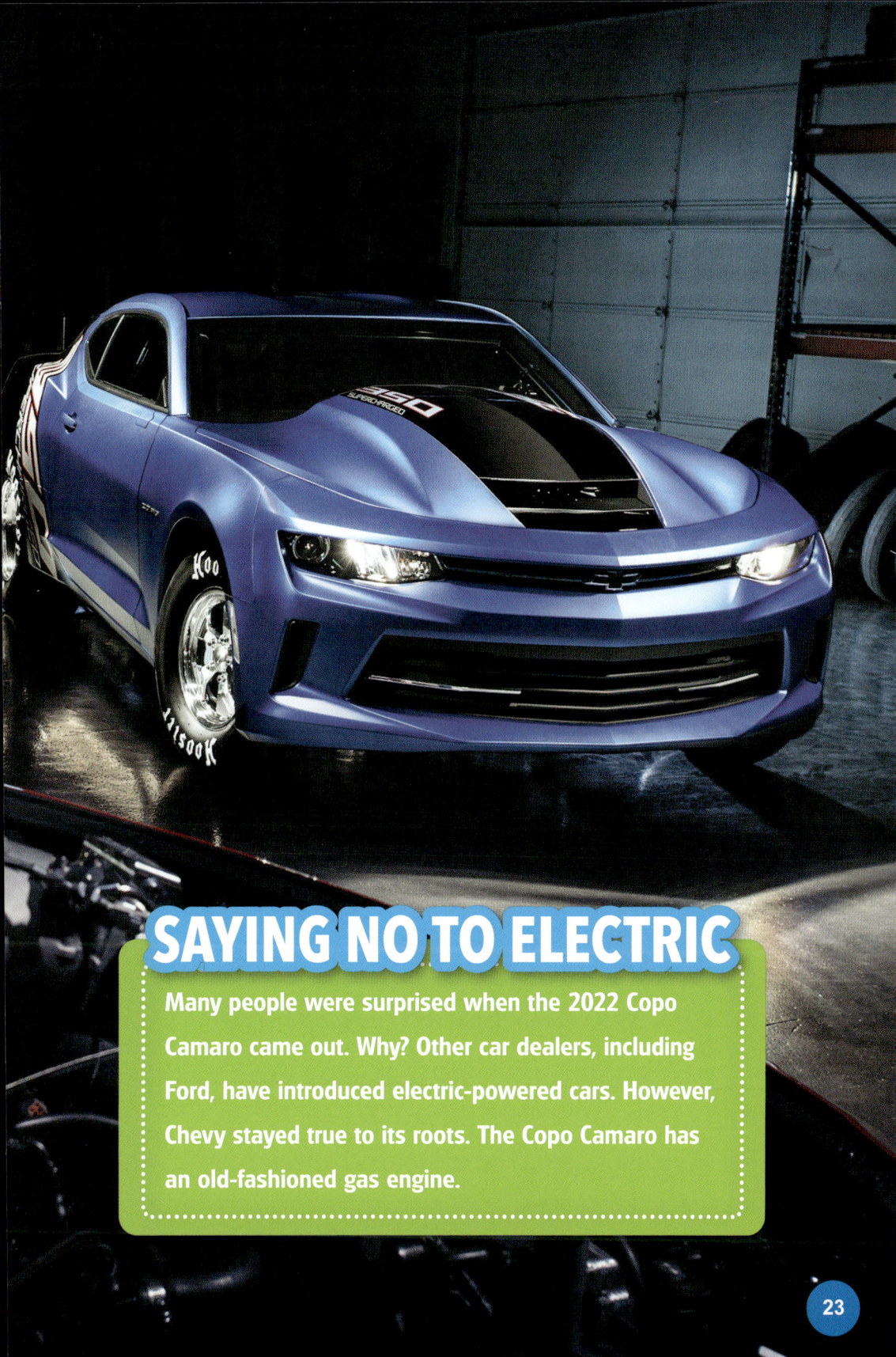

SAYING NO TO ELECTRIC

Many people were surprised when the 2022 Copo Camaro came out. Why? Other car dealers, including Ford, have introduced electric-powered cars. However, Chevy stayed true to its roots. The Copo Camaro has an old-fashioned gas engine.

Chapter 4
Around the Track

Randall eased into the seat of the Chevy Copo Camaro. He adjusted his helmet. Randall was wearing a racing suit to protect him. He needed to be safe because the Camaro would be a very fast ride.

Randall pressed down on the gas pedal. The powerful engine roared into life. The driver sat still for a moment. He enjoyed the throb of the engine. Then he pushed the gas pedal down as hard as he could.

The engine's roar peaked into a high-pitched whine. The tires spun. The back of the car lifted up as the front tires gripped the track. All of this took less than a second. Then the car **accelerated**. It zoomed down the track.

The car was going fast, but Randall knew it could go much faster. He accelerated to increase his speed. The car leaped forward, its engine whining even louder.

FUN FACT
The Copo Camaro appears in National Hot Rod Association Stock Car and Super Stock Car races.

Randall checked the **speedometer**. Its needle touched the 150-mph mark. The needle climbed even higher as Randall hit the straight part of the track.

The car raced over the finish line at 171 miles per hour. Now it was time to slow down. Randall pressed a button. A parachute popped out of the trunk. It caught the air and slowed down the car. In just a few feet, the car was still.

LOOKS CAN FOOL YOU!

Some race cars look very different than the cars you see on the highway. Not the Chevy Copo Camaro. The Copo Camaro is a stock car. It looks like a regular car. However, it has a special engine and other features that make it ready to race on a track.

Randall switched off the engine. He pulled off his helmet. Friends rushed to help him out of the car.

Randall could not stop smiling. "What a great ride!" he said. He could not wait to take another spin in this amazing car.

BEYOND
THE BOOK

After reading the book, it's time to think about what you learned. Try the following exercises to jump-start your ideas.

THINK

DIFFERENT SOURCES. Think about types of sources you could find on the Copo Camaro. What could you find in a magazine? What could you learn at a dealership? How could each of the sources be useful in its own way?

CREATE

GET ARTISTIC. Cars start with creative artists and designers. Time for you to take a shot! Get art materials and create a great, new car. Will you make it a sports car? A sedan? A race car? What colors will you paint it? What features can you give it? Let your imagination go for a spin!

SHARE

SUM IT UP. Write one paragraph summarizing the important points from this book. Make sure it's in your own words. Don't just copy what is in the text. Share the paragraph with a classmate. Does your classmate have any comments about the summary? Do they have additional questions about the Chevy Copo Camaro?

GROW

GO TO A CAR SHOW. Car shows are a great way to see lots of cool cars up-close. Check your local events calendar, or ask at a car dealer for upcoming events. You can find shows of old cars and new cars, sports cars and classic cars. Go to a show and find a new favorite car to love!

RESEARCH NINJA

Visit *www.ninjaresearcher.com/6105* to learn how to take your research skills and book report writing to the next level!

RESEARCH

DIGITAL LITERACY TOOLS

SEARCH LIKE A PRO
Learn about how to use search engines to find useful websites.

FACT OR FAKE?
Discover how you can tell a trusted website from an untrustworthy resource.

TEXT DETECTIVE
Explore how to zero in on the information you need most.

SHOW YOUR WORK
Research responsibly—learn how to cite sources.

WRITE

GET TO THE POINT
Learn how to express your main ideas.

PLAN OF ATTACK
Learn prewriting exercises and create an outline.

DOWNLOADABLE REPORT FORMS

Further Resources

BOOKS

Fishman, Jon M. *Cool Stock Cars*. Minneapolis, MN: Lerner Publications, 2019.

Mason, Paul. *American Supercars: Dodge, Chevrolet, Ford*. New York: PowerKids Press, 2019.

Rusick, Jessica. *Chevrolet Camaro*. Minneapolis, MN: Big Buddy Books, 2021.

WEBSITES

Factsurfer.com gives you a safe, fun way to find more information.

1. Go to www.factsurfer.com.

2. Enter "Chevrolet COPO Camaro" into the search box and click 🔍

3. Select your book cover to see a list of related websites.

Glossary

accelerate: to accelerate means to go faster. The Chevy Copo Camaro can accelerate from 0 to 60 miles per hour (97 kph) in 5.4 seconds.

carbon fiber: carbon fiber is a very strong, lightweight material. Using carbon fiber to build a car makes it lighter and faster.

cubic inch: engines are measured in cubic inches. The cubic inches of a car engine measures the amount of volume in the car's cylinders.

drag racing: drag racing is a race between two or more cars over a short distance. A drag racing track is usually only a quarter-mile long.

horsepower: horsepower measures the power of the engine. The Chevy Copo Camaro has more than 650 horsepower.

speedometer: a speedometer is a display on the dashboard that tells how fast a vehicle is going. The driver saw on the speedometer that he was going too fast.

transmission: the transmission is the part of the car that moves power from the engine to the wheels. The Chevy Copo Camaro has a three-speed transmission.

V8: a V8 engine has 8 cylinders in the shape of a V. The Chevy Copo Camaro has a V8 engine.

Index

carbon-fiber, 5, 10
Chevy, 5, 9, 10, 12, 14, 15, 16, 17, 18
General Motors, 12
horsepower, 16, 18
parachute, 11, 26
racetrack, 9, 12, 14, 15
racing, 4, 10, 11, 16, 21, 24
super-charged, 16, 18
transmission, 14, 21,
V8, 4, 16, 18,

PHOTO CREDITS

The images in this book are reproduced through: Chevrolet Pressroom/General Motors (Vanderkaay/GM 30).
Cover: Courtesy of Chevrolet Pressroom/General Motors, YIUCHEUNG/Shutterstock (background).

About the Author

Joanne Mattern has written many nonfiction books for children. Her favorite topics include sports, biographies, animals, and history. Joanne lives in New York State with her family and loves to drive fast cars.